PLANT PROPAGATION 101

A Practical Guide In Plant Propagation

By

May Alsisto Castillo

Edited by Zel Cabigting
ISBN:
Hardbound-978-621-470-208-4
MOBI/KINDLE-978-621-470-209-1
Softbound/Paperback-978-621-470-210-7

Published by:
Poetry Planet Book Publishing House
Rosario, Pozorrubio, Pangasinan, Philippines
Contact No.: 09554960044
Email: maritesritumalta@gmail.com

INTRODUCTION

Plant propagation nowadays is starting to gain popularity not only among nature protectors, plant lovers, agriculturists, horticulturists, agronomists, and the like but for the entire population as well.

Our planet Earth is trying to thrive and survive a seeming destruction and decay. Admittedly, its condition affects us in many ways whether we are among those who care about our planet or among those indifferent towards it.

Earthquakes, wars, oil spills, flash floods, increase in population, housing and building projects, increased greediness and selfishness of humans, and many others are factors contributing to great loss of greens and even extinctions of them.

Geologist and a member of Science magazine, Erik Stokstad, Quoted: “Twice as many plants had gone extinct than birds, mammals and amphibians combined and an overwhelmingly 600 species of plant have vanished from the wild in the past 250 years.

Earth-wide, campaigns for planting trees and plants, green recoveries, plant propagations, and so on, are among the heroic acts done on behalf of our planet and the entire human race.

The author of this book would like to contribute to these life and earth-saving acts by collecting facts, sharing how-to-do ways, and even encouraging individuals globally to offer their share for our planet.

Please have some time to read and meditate on the information and lessons penned in this book.

TABLE OF CONTENTS

https://www.thekidsgarden.co.uk/around-the-garden/exploring-plant-propagation-with-your-kids/

PLANT PROPAGATION

Each year, more and more plants are going extinct. Today's generation is confined to only a few known species and we will not wait until those few species will even disappear for the incoming next generation.

Plant Propagating will be among the wonderful acts any human can be a part of so that plants can multiply, reproduce and continue to survive. Therefore, plant species and varieties will continue to augment and will even be expanded and a great possibility of being rediscovered. However, there are more benefits acquired from plant propagation. Are you among the so-called *Plantitas or Plantitos*, termed for plant lovers in the Philippines? Are you fond of landscaping your own garden or even your miniature garden during this pandemic? Despite the high cost of plants nowadays, do you still yearn to collect varieties of them just like what your friends and others have? Moreover, maybe you are a herbalist, and you wanted to keep and collect herbs that heal or even protect you, your family, and friends from different diseases. Having those pointers in mind, we may want to add to our hobbies or probably to our chores, plant propagating.

By the way, what is plant propagation? Another description given is the science and art of reproducing plants. It is the production of offspring or multiplying plants by any process of reproduction from a parent stock or increasing their number by sexual or asexual means.

The schools had integrated plant propagating as a part of their curriculum categorized under EPP/TLE for countless years. In addition, during the pandemic when diseases spread so quickly and people were restricted from going out of their houses most of the time, many came to realize its personal benefit for themselves and for their families. The presence of plants in their own backyards provided them food and comfort to a great deal. Therefore, we should not only assume that plant propagating is only a hobby but an essential for survival. It is also important to pass on its benefits and the knowledge to younger generations if we want the earth to keep thriving.
However, there are questions that we need to consider in propagating plants:

1. What tools and equipment do we need?
2. What are the different ways of plant propagation?
3. What challenges are encountered in propagating plants but what solutions to those challenges are available?

The author of this book would like to call the attention of concerned citizens. Why? Because the healthy cycle of life for generations to come continues, plant propagation is among the sources to protect and support ourselves. It is also important to pass on to younger minds the processes involved in plant propagation.

Nevertheless, before we are going to consider the answers to the questions posed earlier, let us compare different ways of propagating from the past to the present.

PROPAGATING PLANT HISTORY

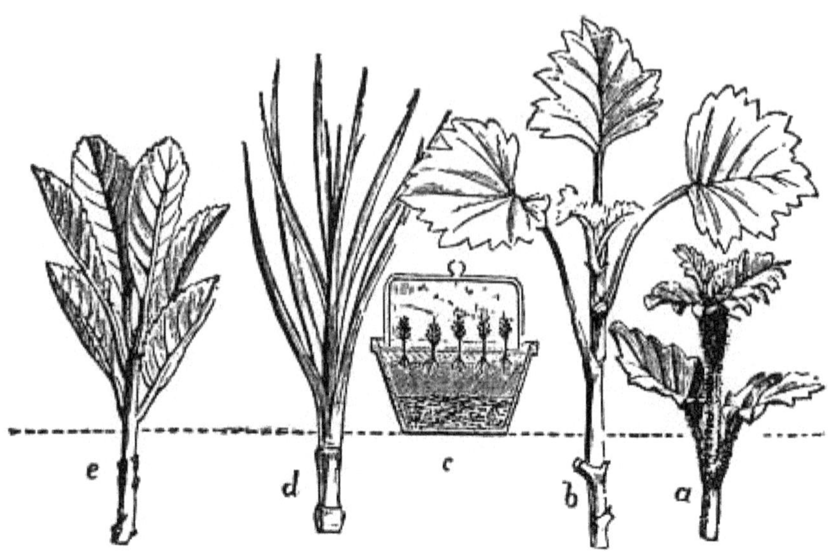

https://en.wikisource.org/wiki/Page:EB1911_-_Volume_13.djvu/774 ...

From Creation Onwards

From creation onwards, humans kept alive when they continually take in fresh air, healthy food, and clean water. Not only do humans have such needs, but animals of all kinds also live with those basic supplies.

Both humans and animals shelter in places producing ample amounts of those supplies then moved to new locations or areas when their place becomes exhausted with those supplies.

However, when people find a fertile land available for them and their domestic animals, they usually decide to stay for a longer period or even the longest period of their lives. Then planting and propagating becomes a part of their routine because their survival largely depends on them.

As the human race increase in number, agricultural areas to propagate also become inadequate. Improved land transportations also affect propagating areas, not only taking a huge portion but they also spread pollution or deadly chemicals that ruin the health and life of plants. Another factor affecting plant propagation is the advancement in technology. Despite efforts to provide for better equipment in plant propagation, the interest of many in this aspect of occupation diminishes as people train to become skillful in the different industries that become available.

On the brink of extinction though, many came to their senses and realized that humans were originally fashioned with the need for the basic supplies of fresh air, healthy food, and clean water. However, the uncontrolled selfishness and greediness of humans produced an inadequate supply of these. Nevertheless, many resorted to adapting modern ways of propagating plants in order for humans to enjoy the beauty of plants and of course, to survive.

10,000 Years Ago

10,000 years ago, when the so-called hunter-gatherer lifestyle halted, then the birth of plant propagation took its toll. Asexual propagation is described to be the production of a genetically identical plant to the parent plant focused on onions, sugar cane, bananas, potatoes, and pineapples during those times.

4,000 Years Ago

Then around 4,000 years ago, The Romans were the first to start grafting woody plants using approach grafting. Approach Grafting is when a plant graft is made by joining stock and scion laterally at an intermediate point but leaving both rooted and uncut until firm union is established when the stock is cut above and the scion below the union.

18th Century

During the entry of modern years, countries between Japan, China, Australia, and the tropics produced an explosion of plant propagating methods, which included the asexual plant propagation. Walk-in greenhouses were then invented and constructed enabling more diverse methods of propagation of being discovered.

1950's

By the 1950s, the invention of a modern intermittent mist system sprinkling a low amount of water every few minutes all day long throughout the day automatically through a system function continually until the cuttings become rooted.

During those years, plastic film was developed. Polyethylene was the first plastic film they used to cover wood frame greenhouses, they also use copolymer poly. The process helps create a microclimate when roots have not yet developed in the first weeks of the root. Either the plastic film be used directly on top of the film or one can opt to make a small tunnel.

Another modern system of propagating called micropropagation is the clonal propagation of plants in closed vessels under aseptic or sanitary conditions. In vitro, which means glass micropropagation is when the plants grow on culture media containing nutrients and growth regulators. On the other hand, in vivo is the description given to plants grown in soil.

1980's to the Present

During the 1980s, Fog system, which consists of incorporating a large number of micro-particles of water into the ambient air, which remained suspended in the air inside the greenhouse for as long enough to allow it to be able to evaporate inside the greenhouse without wetting the crops. Then, by using special nozzles, the distribution of water in the form of fog over the surface of the greenhouse uniformly takes place. In order to humidify and cool down the greenhouse in a controlled manner and in carrying out disinfection treatments using soluble plant protection products, the fog system is very useful.

They developed insect Growth Regulators also known as IGR, and Bottom Heat was used to propagate plants by keeping the top of the cutting supine and instigate growth of roots at the rudimental end. Sanitation, which is an important part of preventing plant propagation diseases, also played an important role in propagating victories.

In the following subtitles, we will be discussing the different ways of propagating plants. First, though, we need to know which equipment and tools are needed to propagate plants in order for us to be equipped while going along with your reading of this book. We would not want to miss the opportunity of learning hands-on and the chance of discovering the act of supporting life.

PROPAGATING TOOLS AND EQUIPMENTS

Bare hands and brute strength are, of course, important tools but to prevent one from acquiring injuries and some minor accidents, the right tools and equipments can help make the job done with lesser friction and to avoid possible minor accidents.

Here are some of the tools we can acquire online or locally:

PROPAGATION/ GRAFTING KNIFE -For good grafting results, a sharp knife is an essential tool. Grafting knives differ from other knives. They have thin, sharp razor-like blades filed on only one side allowing it to cut through tough, woody material easily. For those who do a lot of propagation work, a quality-grafting knife is a very helpful tool.

SHARPENING STONES - Sharpening stones also known as a whetstone. They are simple, longstanding tools for sharpening blades.

HAND PRUNERS – in propagating, these are used to cut smoothly thin pieces of wood or soft parts or stems of the plant.

DIBBLE- used for transplanting purposes, for making holes in the soil, and for planting bulbs and seeds.

GRAFTING CHISSEL AND MALLET- used for trunk grafts and or grafts on larger branches. It looks like a sharp nail hence it is usually used with the aid of a mallet.

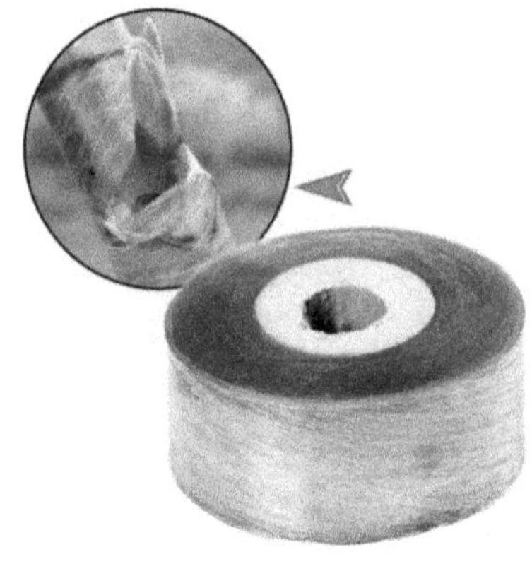

GRAFTING WRAP OR TAPE- enables the protection and consolidation of the graft and callus because of microcrystalline wax impregnated in the cotton tape. It is also useful in reinforcing the unity between the scion and rootstock, thus protecting it from the air and water causing dehydration. The Indugref brand is used for all types of plants and grafting methods.

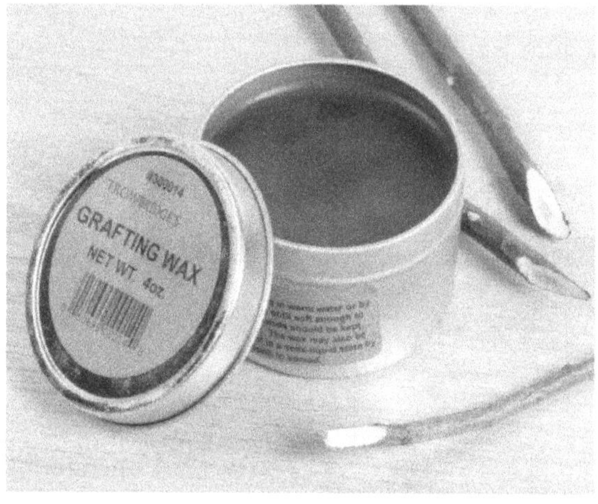

GRAFTING WAX- is a composition of beeswax, rosin, tallow, and similar materials, used in gluing and sealing the wounds of newly grafted trees or shrubs to protect them from infection.

LABELS-or naming the propagated plant is important. It should be clear and specific. Make sure that the labels will keep its clarity. It can help to put seals or top-ups that will prevent it from becoming misty. Some details in labels containing the date, plant name, and the name of the collector may be included.

SIEVES- can be at any desired shape, that is because its main function lies in its holes. The sieve is used to separate large particles of soil from finer particles, which would pass through the holes of the sieve. The passing soil particles would then, be used to take the role of caring for seeds until they germinate.

POTTING BENCH- can be made of wood or any material available as long as it possesses a sturdy base. Wheels or the number of shelves depends on an individual's choice and preference.

CONTAINERS- plastic pots and blocks, paper pots, polythene pots, peat pots, soil and peat blocks, bedding plant pots, trays, seed boxes, and propagators are among the lists of possible containers whichever is available and depending on the type of propagating option.

HORMONE ROOTING COMPOUND - since they contain cytokines, that is another plant growth hormone, also fungicides and some other chemicals, it reduces the risk of plants succumbing to fungal infections. In addition, the chances of your cuttings to take root increase.

HEATING PAD- a covering used for maintaining a set temperature to make sure that the plant seeds grow quickly. Due to the many changes in the environment, using a heating pad accelerates the growth of plants, as well as confirms that the plants grow healthy.

PROTECTED AND CONTROLLED ENVIRONMENT – This is a space or area that has many different forms in which plantations can be closely monitored and cared for, which is an effective method to reduce the threat of diseases and pests, maximize efficiency, increase sustainability and yield, and also to bring down the overall cost of operation. Indoor, greenhouse, vertical farming, protected cropping, are different kinds of growing environment

Even with the lists of these tools and equipment available, others opted for improvised tools that are low in cost, available, and handy, and they proved useful especially for plant propagation in miniature scenarios. By having these lists in mind, one can initiate the art of plant propagation on their own. Moreover, as they expand in interest, these lists of tools and equipment would be handy for them to recognize and fill out their needs to succeed in their endeavor in plant propagating.

Now that we already considered Tools and equipments for propagating plants, it is important for us to know how plants propagate. Please read on.

2 TYPES OF PLANT PROPAGATION

We must keep in mind that most plants have the capacity to propagate both sexual and asexually enabling plants to evolve, adapt, and also to colonize and dominate new territory quickly.

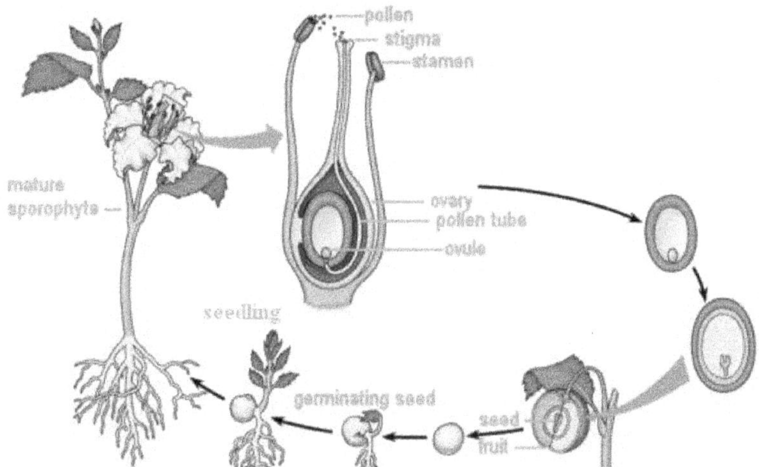

https://sites.google.com/site/meride592/home/introduction/definition/types-of-propagation/sexual-propagation

SEXUAL PROPAGATION

This describes the reproduction of plants by seeds. By pollination and fertilization that combined the genetic material of two parents, they produce offspring that are different from each other. Sexual propagation is advantageous are, it is quicker and more economical for propagating. It also produces amazing results in new cultivars and vigorous hybrids.

It also minimizes transmissions of particular diseases, like viruses. The plants' genetic variations are also maintained by this type of propagation thereby increasing the potential for plants to adapt to environmental pressures. Another advantage of sexual propagation is that, for some plants, this is the only way for them to propagate.

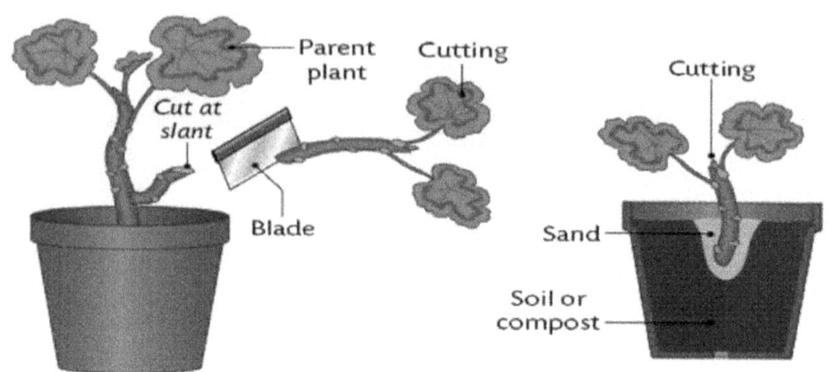

https://classnotes.org.in/class7/science-7/reproduction-in-plants/asexual-reproduction-in-plants/

ASEXUAL PROPAGATION

Asexual propagation sometimes is referred to as vegetative propagation because some parts of the plants like stems, roots, or leaves are taken to regenerate into a new plant or, intended to result in several plants. Of course, by only a few exceptions, the resulting plant is usually genetically identical to the parent plant.

This type of propagation involves cuttings, layering, separation, division, grafting, micropropagation, and budding. For some species, asexual propagation can be easier and faster compared to sexual propagation. It also maintains juvenile or adult characteristics of certain cultivars and for some cultivars, this type of propagation may be the only way to perpetuate them. This also allows the propagation of special types of growth like weeping or pendulous forms. Asexual propagating also results in producing quickly large plants.

Now that we have discussed how plants increase or reproduced through sexual and asexual propagation, we can now choose ways of propagating them either through the sexual or asexual propagating procedure. Please enjoy discovering different ways of propagating and be refreshed by the fact that despite all your efforts and sacrifices, you will not only contribute to a greener planet earth but you will also find satisfaction for being able to learn beautiful yet life-saving hobbies.

DIFFERENT METHODS OF GROWING AND INCREASING PLANTS

SEXUAL PROPAGATING METHODS

SEEDS

High-quality seeds can have an impact on the plant produced from it, so it is important to choose seeds of high quality. It would be searched for them from a reliable source, like LGUs that promote propagating, experienced seed propagators themselves, and some stores offering a high quality of seed though they may cost higher compared to others. It is essential to select cultivars that provide the desired size, color, and growth habits and they should adapt to your area. Always remember that hybrids have more vigor, more uniformity, and better growths than no hybrids although they cost more.

Purchase only enough seed to avoid the likelihood of the decrease in germination. However, should there be more than enough seeds purchased, they must be laminated or foiled, or tightly sealed, then store in a maintained 40 degrees Fahrenheit in low humidity.

Most gardeners save seeds from their own plants especially when they are from hybrid plants. The seeds harvested can be placed in an airtight jar and refrigerated for future use.

GERMINATION

After a dormant period, there are three important factors that need to be observed first before resuming the growth of an active embryo of a seed. First, the seed must be viable, and its embryo must be alive, hence capable of germination. Second, so that it will be favorable for germination, make sure to check its internal conditions. Moreover, you must eliminate any barriers to its germination, in its physical, chemical, or physiological aspects. Lastly, the provision of moisture from water, the right temperature, oxygen, and light for some species are important because they are appropriate for growth.

When germinating a seed, it is important to make sure that an adequate and continuous supply of moisture them since germinating seeds absorb and hydrate water. Without any moisture to absorb, the embryo may die. As was advised, for some seeds to germinate, an ample amount of light is important while others prefer to germinate in the dark. These important details can be read from the instructions provided when buying the seed.

In providing oxygen during germination, make sure that the soil is loose or will aerate, or else the germination will delay or retard. Please make sure to follow instructions regarding the needed temperature percentage for your choice of seed to germinate because not all seeds react to the same amount of temperature.

SEED DORMANCY

These feasible seeds do not germinate but they can be regulated either by the environment or the seed by itself. Its dormancy is caused possibly by some unfavorable conditions and inappropriate supply of its propagating requirements including oxygen, light for some species, proper temperature, and sufficient moisture. And there is the probability that the termination of the germinating came from the seed itself through its internal or external source.

In order to break dormancy, seed scarification and seed stratification may help. Seed scarification is the process of breaking, scratching, or mechanically altering the seed coat. This act can allow water and gases to be absorbed. Some opt to soak these seeds in concentrated sulfuric acid from a few minutes to several hours according to the need of the dormant seed, after it is removal from being soaked; the seed is washed and then sown. Another way for scarification of dormant seed is putting it in newly boiled water, soaking until the water cools, and then after removing it from the water, allow the seed to dry.

On the other hand, seed stratification is used for dormant seeds in an environmental condition that is not favorable for their survival. The mixture of seeds and an equal volume of a moist medium in a tightly closed jar that are then stored in a refrigerator is termed as the cold stratification. Keep in check that the medium remained moist only, and never wet. The same process goes with warm stratifications of dormant seeds; only the temperature must be in 68 to 86 degrees Fahrenheit.

For seeds caused by a combination of dormancy, or seeds that have both internal and external issues, administering scarification first followed by stratification before planting the seed may help.

ASEXUAL PROPAGATING METHODS

CUTTINGS

This type of propagation involves the rooting or producing new plants from a severed part of a plant. For successful propagation by cuttings, the environment of a greenhouse is necessary. However, in rooting a few cuttings only of a plant, a small flat flowerpot may be useful enough, just make sure to maintain high humidity. Placing the container into a clear plastic bag may help induce humidity.

It is very important to create holes in the bottoms to act as its drainage. While for a more elaborate structure, an intermittent mist system is usually required. Pay attention to the quality of the rooting medium. It must be sterile, low in fertility, and well-drained. These factors will help provide sufficient aeration and moisture.

The different types of stem cuttings are ***Softwood***, from the first flush of new growth, ***Greenwood or Herbaceous,*** young stems, ***Semi-ripe,*** developed buds, ***Hardwood,*** from dormant wood. As for the leaf cuttings, the choice of ***Leaf bud*** from semi-ripe stems usually trailing vines, with a leaf and an axillary bud, ***Leaf-cutting*** is the whole leaves or section of a leaf. ***Root Cuttings*** is taking the strong long, healthy roots that are 2-3 years old during the dormant season while the level of carbohydrates is high.

Some important points to keep in mind with your choice of plant for cutting must currently from the seasons' growth, remove the flowers, plants must be disease-free, and it must not be under water stress. Therefore, cuttings taken from young plants and from lateral shoots are better than by those taken from older plants and from terminal shoots.

SEPARATION AND DIVISIONS

This type of asexual propagating is one of the easiest and quickest methods. This is the separating of some plants into several self-supporting new batches. The process involves digging up the plant or even removing it from its container then cutting or dividing it, separating them into pieces. Most plants responding to division propagations are herbaceous perennial plants and some woody shrubs. Division plant propagation is usually done during the early spring or in the fall season.

LAYERING

This is when there is a development of roots on a stem while still attached to the parent plant. When the rooted stem is separated from its parent plant, it can be propagated naturally but still needs assistance in the process because the availability of aeration and constant supply of moisture is still important to its success. Though it may also be noted that when rooted stems came in contact with a rooting medium, a natural propagation usually occurs.

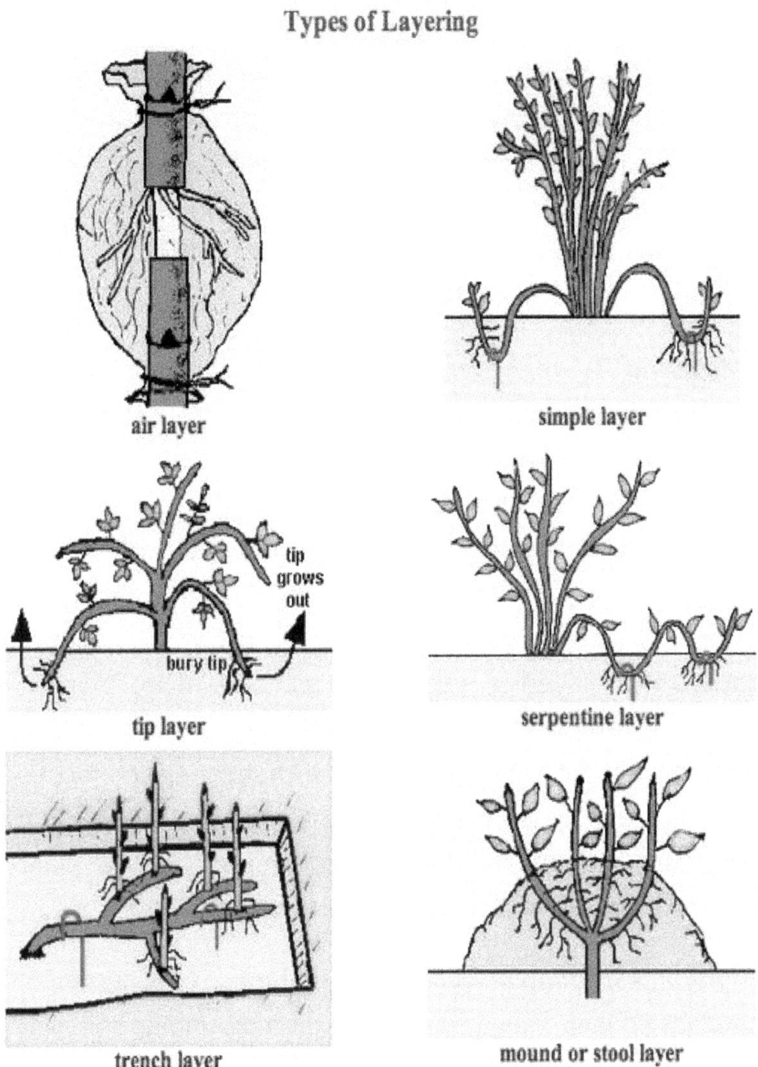

air layer
simple layer
tip layer
serpentine layer
trench layer
mound or stool layer

greengardenblog.com

There are various types of layering. ***Self-layering*** is the contact made by a rooted stem of a parent plant to a rooting medium nearby. ***Tip-layering*** is done by inserting the tip of the cut shoot into a 2-3 inches hole then covering it with soil. Expect the growth downward then it would bend sharply or recurve to grow upward. Another is the so-called ***Simple layering,*** which is done by bending the stem to the ground giving its lower side a sharp bend while covering a part of it with soil. Plants with low-growing branches can be mediums for this type of layering.

Air-layering is for overgrown houseplants and woody ornamentals like the camellias. Monocots like corn plants must be cut upward one-third of the way through the stem holding it open with a wooden matchstick or a toothpick, dusting it with rooting hormone surrounded with damp, unmilled sphagnum moss then wrapping it with aluminum foil after which taping it so as not to be moved. Dicots' process of air layering is much the same as monocots. But for dicots, remove 1-inch ring of its bark and then follow through with the way it was done with layering monocots. Sever the stem below the medium after the rooting medium filled itself with roots. Be ready to pamper the new plant until the root system becomes more developed.

Traditional Stooling or Mound layering meant the cutting of the plant an inch above the ground then mounding soil over the new shoots while waiting for them to grow. By the bases of the young shoots, its roots will then develop. This is being accomplished during the dormant season. Remember to remove the layers on this season of dormancy.

French or Trench –Layering is burying the whole stem, then after some time, digging and dividing it. As for plants producing vine-like growth such as heart-leaf philodendron, pothos, and grapes, bending their stems to the rooting mediums while alternately covering and exposing stem sections and also wounding the stems lower side is called ***Compound or the serpentine layering.***

Natural layering is the type that does not require the help of a propagator. Plants producing runners are severed and even rooted from their parent plant and placed in a different rooting medium, thus developing their own roots and so goes its natural cycle.

STORAGE ORGANS

This acts as a storage house for nutrients and plays a vital part in the survival of a plant. They are buried underground during their growth process while storing carbohydrates and water to keep the plant living longer and provide backup supply of nourishment until the harvest.

Bulb is formed from the meeting of the leaves and the stem. Amazingly, it can withstand weather, temperatures, and seasons because of the plentiful layers, keeping it safe. Most of those layers store the food and hold energy during the growing season.

Root Tubers do not have scales and leaves, but they are commonly enlarged, round, and fleshy organs developing from a stem or root. They only form shoots at their crown. They also store very well when food is scarce during the winter months.

Another underground storage organ, formed from the stem of a plant, is the ***Corm.*** They are useful for plants to survive harsh conditions including fire and cold weather, too. Plants grow from them under favorable conditions.

Next are the ***Rhizomes***, plants that spread their roots, which grow along the ground though they do not have recognizable stems and leaves. They are even able to grow under other plants however, they die off as they age and do not produce annually.

	Plant	Storage Organ
A	Irish potato	Bulb
B	Onion	Stem tuber
C	Cassava	Root tuber
D	Carrot	Rhizome

GRAFTING AND BUDDING

This is a skill, a science, and an art of joining two or more different plants together and when they grow, they reunite as one plant. This method is being reserved for plants that do not root from cuttings yet make its fruiting and flowering mature on a faster level, it can offer resistance from disease, and control the scion's top growth size.

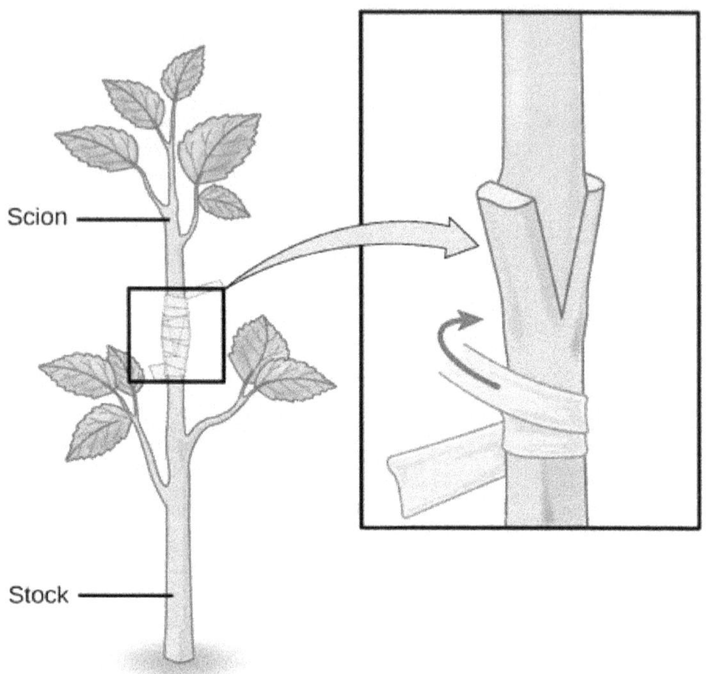

There are six factors to consider to succeeding in grafting and budding. First, is the compatibility of the rootstock and the scion. They must be at the proper physiological stage of growth is the second. Third, to consider is their cambial region, make sure that they are in close contact, even touching one another. Proper polarity, meaning the buds and scion are pointed upward should be maintained. Fifth, protect all cut surfaces immediately after grafting from drying out and the sixth is the proper care given for the newly grafted graft.

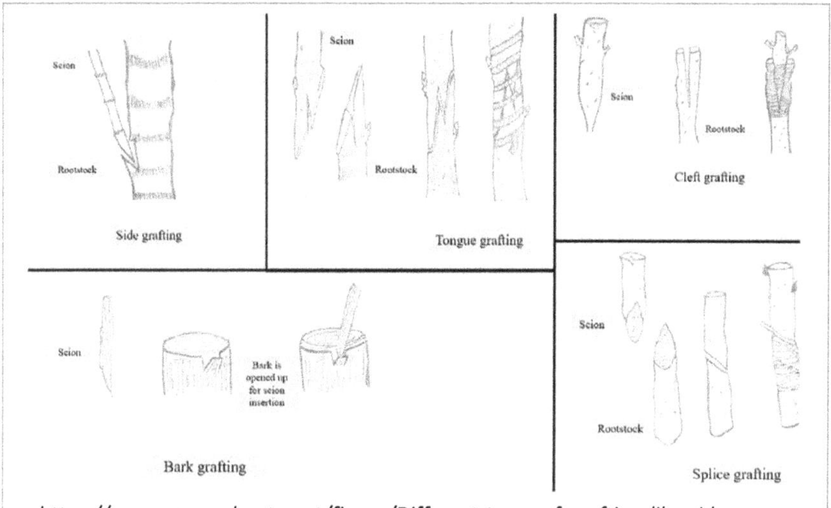

https://www.researchgate.net/figure/Different-types-of-grafting-like-side-tongue-cleft-bark-and-splice_fig1_346925092

Some types of grafting are ***Approach grafting,*** where roots kept attached. However, roots that are cut off are called ***Detached grafting. Whip and tongue grafting*** is the interlocking of the scion and the rootstock together then tying or waxing the wrapped union.

The oldest and most widely used grafting though is the ***Cleft grafting.*** This type of grafting unites the much larger rootstock limb to the scion piece. ***Splice grafting, side, stub,***

side veneer, and side tongue almost portray the same procedures; each has advantages for special situations.

Budding methods on the other hand consists of ***T-Budding,*** a most commonly used technique in propagating many fruit trees. ***Chip budding*** is used when the bark is not slipping. Wrapping it with polyethylene tape will play an important role in avoiding the bud from drying thus forming a miniature plastic greenhouse over the healing graft.

Taking proper care of the plants for a year or two after budding and grafting are where the success of this propagating depends. Different ways of caring for the plants according to the type and methods used in grafting and budding are essential.

After having learned some basic and different types of propagating plants via sexual and asexual propagation, it is important to thoughtfully consider other factors involved in this project. The following parts of this book will give some important information that would contribute to the success of plant propagation.

SOILS AND MEDIA

The soil is like a storehouse for all the elements including nutrients, organic matter, air, and water that plants need in order to grow. It also supports the roots of a plant. However, when left uncared, the soil will only be fit for growing weeds.

Sandy soils could not hold enough water while clay soils hold too much water and do not allow enough air to enter the soil. Plant materials, manure, compost, sawdust, green manure, are organic matters that make soil more workable.

For starting plant cuttings, a soilless media is the best starting mix. Make sure that the mixture is loose, has plenty of oxygen, and well-draining.

Growing media includes peat, coir pith, wood fibers, bark, composted materials, mineral constituents, and are formulated from blending and enriching fertilizers, lime, and some biological additives. Growing media is as important as water and fertilizers for optimal plant growth.

CONCLUSION

After taking a tour of these plant-propagating projects, you would have probably thought of starting a very exciting journey with the greens that provide food and fresh air conducive to a calm, happy life.

Please take one step at a time and at the same time do not procrastinate on this wonderful project. Always remember that it will not only become a joyful hobby nor a burdensome chore but a fulfilling responsibility to save our planet, the generations to come, and our wonderful future.

Let us make plant propagation a part of our system, ensuring the cycle of life and hope for everyone. Please Enjoy Plant Propagating!

BIBLIOGRAPHY

Science web article June 10, 2019 by Erik Stokstad

Plant Propagation 101 by Adam Blalock

General aspects of Propagation –part one- Aggie Holticulture

https://aggie-horticulture.tamu.edu>faculty>davies

Mike McGroarty, Propagation Using Intermittent Mist to Root Your Cuttings, May 13, 2015

Curriculum Resources for Michigan Agriculture Teachers Segment 8. Plant Culture & Propagation, Standard: Demonstrate plant propagation techniques.(Technical II.C.1)

Greenhouse Management .com May 2010, Columns – Technology, John W. Bartok Jr.

ULMA
ulmaagricultura.com/en/greenhouses/equipment/fog-system/

A.v. Roberts, A. Schum, in Encyclopedia of Rose Science, 2003

SFGATE Home Guides /Garden /Gardening by Julie Richards

GrowerTalks, Preventing Disease in Propagation by Tami Van Gaal, March 3, 2016

eHow, How to Use Grafting Knife by Frank Whittermore

thespruce.com/how-to-use-pruners-secateurs-3269518 By Jonathan Landsman, updated 06/02/19

Gardening Info Zone, gardeninginfozone.com/containers-for-propation-and-seed-sowing, March 2011 by Dave Pinkney

Gardeners' World.com, by BBC Gardeners' World Magazine

Pinduoduo, stories.pinduoduo-global.com/agritech-hub/controlled-environment-agriculture, External Guest Writer, April 27,2021

Extention University of Missourri, Missouri Master Gardener Core Manual, David Trinklein, Division of Plant Sciences

Wax and Grafts, waxandgrafts.com/grafting-tapes#

Wikipedia The Free Encyclopedia, en.wikipedia.org

North Carolina Extension Gardener Handbook,content.ces.nscu.edu/extension-gardener-handbook/13-propagation#::-:text=Sexual%propagation

Build a Stash, buildastash.com/post/what-are–food-storage-organs#,Mark Walker, updated September 2,2021

Plant Propagation Cleft Grafting, aggie-horticulture.tamu.edu

Read more at Gardening Know How: Starting Plant Cuttings – How To Root Cuttings From Plants https://www.gardeningknowhow.com/garden-how-to/projects/rooting-plant-cuttings.htm

Texas A&M Agrilife Extension, Soil preparation by Joseph Masabni,
https://agrilifeextension.tamu.edu/library/gardening/soil-preparation/

ABOUT THE AUTHOR

May Alsisto Castillo is a member of the Federation Of teachers' Association, like all other mothers, loves cooking and gardening. She is a teacher by profession with special skills of driving and knowledge of computer operations, and she has also expanded her career by joining seminars, webinars, and training that enhances her knowledge, skills, abilities, and even her own personality.

During the year 2020, she had been busy, not only with keeping her career a priority, but she also attended Division Webinar in Physical Education and Health 2020, Division Online Training On Core IT Skills (AKT-2ND PHASE), Division

Webinars Skills Training For EPP/TLE/TVL Teachers, Webinar sessions On Covid -19, Division Orientation On Most Essential Learning Competencies, Division Webinar in Video Making, and twice on the Division Inset 2020.

At the entry of the year 2021, she kept herself updated by being active with more trainings provided to teachers. She has joined Oplan Kalusugan Sa Deped One health Week Virtual Kick-off, Oplan KAlusugan – Oral Symposium: Oral Health During Pandemic, Medical, and Dental services e-Health Technology Meets Healthcare In The Department Of Education, Wash in school Implementation During The Pandemic and Health Education, School-Based Feeding Program During The Pandemic – Launching Of Virtual Cooking Show for Learners, And The National Drug Education Program Orientation and Eskwela Ban sa Sigarilyo.

Moreover, during the start of this year of 2022, she is unstoppable. She keeps looking for opportunities to expand herself even more. One of the components of TLE and EPP is entrepreneurship and ICT. That is why at the beginning of this 2020, She embarked on an Income Generating Project which is called, **"GRAFTING AND SEEDING PROPAGATION OF DESERT ROSE".**

As she keeps stepping up on a wonderful journey and wonderful adventures not only for herself but also for her loved ones, friends, co-workers, and students, the start of this year moved her to author this book, **"PLANT PROPAGATION 101: a Practical Guide In Plant**

Propagation". More years will come after this and May will keep on growing, expanding, and shining the best way she can.

www.ingramcontent.com/pod-product-compliance
Ingram Content Group UK Ltd.
Pitfield, Milton Keynes, MK11 3LW, UK
UKHW022007190726
13853UKWH00004B/1794

9 786214 702107